The Story of Science

Dance of the Continents

by Roy A. Gallant

BENCHMARK BOOKS

MARSHALL CAVENDISH
NEW YORK

Series Editor: Roy A. Gallant

Series Consultants:

LIFE SCIENCES
Dr. Edward J. Kormondy
Chancellor and Professor of Biology (retired)
University of Hawaii—Hilo/West Oahu

PHYSICAL SCIENCES
Dr. Jerry LaSala, Chairman
Department of Physics
University of Southern Maine

Benchmark Books
Marshall Cavendish Corporation
99 White Plains Road
Tarrytown, NY 10591-9001

Library of Congress Cataloging-in-Publication Data
Gallant, Roy A.
        Dance of the continents/ by Roy A. Gallant.
        p. cm. — (The story of science series)
Includes bibliographical references and index.
Summary: Describes the development of geological theory from ancient Greek philosophers to the discovery of plate tectonics, which explains the forming of geological structures.
ISBN 0-7614-0962-9
        1. Plate tectonics—Juvenile literature. 2. Geology—Juvenile literature. [1. Plate tectonics. 2. Geology.] I. Title. II. Series: Gallant, Roy A. The story of science series.
QE511.4.G35   1999   551.1'36—dc21   98-28046   CIP   AC

Photo research by Linda Sykes Picture Research, Hilton Head, SC
Diagrams on pp. 20, 25, 32, 36, 45, 46, 49, 55, 57, 59, 60, 64, 65, 71 by Jeannine L. Dickey
Cover photo: Kraft/Explorer/ Photo Researchers; 6 Roger Ressmeyer/ Corbis; 9 David Hardy/ Photo Researchers; 10 Edwin Grosnevor/ National Geographic Image Collection; 11 Royal Society/ e. t. Archive; 12 National Geographic Image Collection; 13 James Stanfield/ National Geographic Image Collection; 14 Bill Eldredge; 18 Jonathan Blair/National Geographic Image Collection; 26 Bibliotheque Nationale, Paris; 26 The British Library, ms. add. 28681, folio 9 (wf); 31 James Amos/ National Geographic Collection; 34 Roy Gallant; 39 Culver Pictures; 40 M. Lustbader/ Photo Researchers; 43 Alfred-Wegener-Institut fur Polar und Meeresforschung; 52 David Hardy/ Photo Researchers; 61 Kevin Schafer/ Peter Arnold

Printed in Hong Kong
6 5 4 3 2 1

For Melissa and Martha

ACKNOWLEDGMENTS

My thanks to long-time friend and geologist Professor Christopher J. Schuberth, Armstrong Atlantic University, Savannah, Georgia, for reviewing the manuscript of this book, especially for his valuable comments on those chapters dealing with Earth's mantle and hot spots. My thanks also to Jeannine L. Dickey for her assistance with research and organizing the numerous publications used as reference materials.

Contents

One

When Earth Roars

Every now and then planet Earth does something so suddenly and with such great force that it sends people screaming into the streets. A severe earthquake that kills tens or hundreds of thousands in only a few minutes is one example. A hundred-foot high ocean wave, called a *tsunami* and created by an undersea earthquake, is another. Such waves have crashed down suddenly onto unprepared victims and washed them and their houses into the sea.

Sometimes the onslaught announces itself, such as the warning rumbles of a volcano soon to erupt. Our mightiest human displays of force pale when compared with the tremendous forces operating in nature. The hydrogen bomb seems only a firecracker when compared with the forces that cause a volcano to erupt or that unleash a major earthquake.

Mount Etna, on the island of Sicily, pours out a river of lava nearly every time it erupts. The mountain's frequent outbursts have been recorded since 1500 B.C. Mount Etna can be expected to continue its eruptions as it slowly slides into the Mediterranean Sea.

Disaster Strikes at Mount Pelée

On May 8, 1902, a few minutes before eight in the morning, 40,000 people were engulfed by a hurricane of flame. The flame swept down the slopes of Mount Pelée, an active volcano overlooking the city of Saint Pierre, on the island of Martinique in the West Indies. In only minutes what had once been a colorful seaside city was reduced to charred ruins by the fury of a raging volcano.

A month before the eruption residents of Saint Pierre had noticed wisps of smoke curling lazily out of the quiet mountain's crater. Hikers climbed the slopes to investigate. When they returned they reported hearing rumblings deep within the mountain. Pelée, they feared, was about to go on a rampage. Others scoffed at the idea. After all, Pelée had been sleeping for fifty-one years. Its reawakening in a rage seemed unlikely.

As the days wore on, the townspeople began to hear muffled explosions within the mountain. Then peaceful Pelée erupted. It began hurling clouds of ash from its crater. Frightened animals moved down

When Mount Pelée erupted on the morning of May 8, 1902, it cast off a raging cloud of flame, ash, and rocks that flowed down onto the town of Saint Pierre on the island of Martinique and killed some 40,000 people. That disaster serves as a reminder of the tremendous forces unleashed from time to time by the activity of Earth's restless interior.

the slopes, away from the noise, ash, and tremors. A few days later the island was coated with a film of white ash. On May 5 there were more explosions, this time louder and stronger. A rain of boiling mud and an artillery of rocks came flying out of the crater. By then, no one doubted that a tragedy was at hand. Many people tried to flee Saint Pierre, but reports have it that the local governor posted soldiers to prevent anyone from leaving. Elections were only a few days away, and the governor needed votes to remain in office.

Meanwhile, several ships entered the harbor and joined

those calmly resting at anchor. Then, on that fateful morning three days later, Pelée, like a giant flamethrower, flung its breath of fire down the mountainside onto the town. Except for one person, all were killed. That one was a prisoner locked in the protective dungeon of the local jail. One eyewitness on a ship anchored out in the harbor reported what he saw:

> The town vanished before our eyes, and then the air grew stifling hot and we were in the thick of it. Wherever the mass of fire struck the sea, the water boiled and sent up vast clouds of steam. I saved my life by running to my stateroom and burying myself in the bedding. The blast of fire from the volcano lasted only a few minutes. It shriveled and set fire to everything it touched. Burning rum ran in streams down every street and out into the sea. Before the volcano burst, the landings of Saint Pierre were crowded with people. After the explosion not one living being was to be seen.

Krakatau and Vesuvius

In 1883 the volcanic island of Krakatau, near Java and Sumatra, blew its top in one of the greatest explosions of modern times. On August 26 the mountain island began erupting in a series of explosions. Then, the following day, there was one mighty burst of flame, smoke, and ash. Clouds of ash were hurled 17 miles (27 kilometers) into the air.

When the waters calmed and the air cleared, there was nothing. The mountain had vanished from the surface of the sea. At one point, where the island had risen 2,600 feet (790 meters), the sea became 900 feet (275 meters) deep. The noise from the explosion was heard 1,700 miles (2,735 kilometers) away in Australia, and a pressure wave of air traveled around the world. The sea, churned up by the explosion, towered in a

This painting shows the Krakatau volcano in an early stage of eruption, after which the island was blown to bits. All that remained was a deep hole in the ocean floor where the island had once been. The noise from the explosion was heard 1,700 miles (2,735 kilometers) away in Australia.

great tsunami wave that raced outward and crashed down on the coastal villages of Java and Sumatra. The wave was 130 feet (40 meters) high and killed some 37,000 people. For more than a year after Krakatau blew up, ash and dust from the explosion hung in the air around the world, causing purple snowfalls and green sunsets.

A similar catastrophe struck on August 24 in the year A.D. 79 on the shores of Italy's Bay of Naples. After a thousand years of sleep, the volcanic Mount Vesuvius awoke and erupted with

The Russian painter Karl Bryullov imagined what the scene might have been like in Pompeii during the devastating eruption of Mount Vesuvius in the year A.D. 79. Hundreds of families were buried alive.

Suffocating clouds of gas and ash from the Mount Vesuvius eruption rained down onto Pompeii and completely buried the city. Many people were overwhelmed and buried where they fell. When archaeologists eventually uncovered the city, they found solid blocks, or molds, of ash encasing twisted bodies and preserving the bodies' shapes. After such a mold was cleaned out, plaster could be poured in. The mold's shell was then broken away to reveal a cast of the victim.

terrible explosions. Torrents of ash completely buried the town of Pompeii, while mud flows overwhelmed the nearby town of Herculaneum. Hundreds of families were buried alive and cemented where they were overtaken by the mud. Roman sentries were buried at their posts and turned into statues.

What caused these and countless other devastating Earth upheavals recorded since ancient times? It was just such events that made people wonder how Earth's interior could trigger such monumental destruction at its surface.

A giant mushroom cloud
of hot ash explodes out
of Redoubt Volcano in
Alaska. In 1989 an
airliner flying 200 miles
(322 kilometers) north of
the volcano lost power
when fine ash interfered
with the plane's engines.

Two

Earth's Shape and Deep Inside

A Flat Earth and a World Ocean

For thousands of years people of many different cultures believed that Earth was flat and shaped like a great drum. Some of the early Greek philosopher–scientists who lived around 500 B.C. thought that Earth stretched off only a small distance around the Mediterranean Sea, where they lived. It spread southward to North Africa and northward to a range of high mountains. To the east was the great "world ocean," and to the west was more land. Quite likely people who lived long before the ancient Greeks also thought that Earth did not continue very far beyond where they lived.

Around 500 B.C. the Greek philosopher Anaxagoras taught that Earth was flat and floated on a cushion of air. He also said that Earth was hollow and that the oceans and rivers all flowed out of great caverns within the planet. Earthquakes, he believed, were caused when air above the planet rushed down and plunged against the air cushion beneath. People of many cultures have believed that Earth must be supported in space by *something*. And whenever its support moved, Earth shook. Certain people ranging from the Pacific Islands to Eastern Europe at various times have supposed that a gigantic buffalo carried Earth on its back. Whenever the beast shifted its weight from one foot to another, there was an earthquake. The Algonquian Indians thought the animal was a giant tortoise. In Mongolia it was a frog; in the Celebes, a hog; in Persia, a crab. In India seven serpents took turns supporting the planet. Each time a new serpent took over, Earth trembled. The Greek philosopher Thales, who lived around 600 B.C., supposed Earth was supported by water as it floated in the great world ocean. Water seen to gush from the ground during large earthquakes was evidence, he said. The geologist L. Don Leet relates the following tale in his book *Causes of Catastrophe*:

> According to the Masawahilis of East Africa, a monster fish called the Chewa swims in the world ocean and carries on its back a stone upon which there is a cow which carries Earth on one of its horns. Whenever the cow shifts Earth from one horn to the other, there is an earthquake.

One of the greatest of the Greek philosophers was Aristotle, born in 384 B.C. His teachings about Earth as a planet were to be honored for nearly two thousand years. He said that Earth was not flat but instead was a sphere. You can see this for yourself, he explained, if you watch a ship sail over the horizon. First the body of the ship drops slowly out of sight, then the masts and sails disappear. Does this not show that the oceans are curved and not flat, he asked? And during an eclipse of the Moon, he further pointed out, Earth's curved shadow cast on the Moon shows that Earth is round, not flat. People who listened to

Aristotle's arguments but were still not convinced asked why, if Earth were round, people on the underside did not "fall off." Or how they managed to walk "upside down." Eventually, the law of gravitation would explain why, but gravity was not to be investigated for more than 1,500 years. Since all things on Earth are attracted toward the planet's center by gravity, there is only one "down," and that is the direction of Earth's center.

Aristotle also attacked earlier beliefs that within Earth were water-filled regions from which rivers and the oceans flow. Rivers, he informed his students, were formed in the mountains and eventually flowed into the sea. He also taught that "land and sea exchange places and one area does not always remain land, another sea, for all time, but where there is now sea there is at another time land." This accurate observation later became an important principle in the history of geology.

The Causes of Earthquakes

When he tried to explain what caused earthquakes, Aristotle "launched a thousand ships of fantasy," according to Leet. He taught that winds blowing into caverns within the planet became trapped. When the pressure of the trapped air became strong enough, the air escaped forcefully and caused Earth to tremble. Earlier, Democritus had supposed that hollows within Earth collected water, and when the water sloshed back and forth, the planet shook. Time after time the early Greek thinkers spoke of caverns and great hollows within Earth. Possibly the many sea caves found throughout the Greek Islands gave them that idea.

Ghosts of some of Aristotle's beliefs haunt us to this day. One is his notion of "earthquake weather." Many people think, incorrectly, that humid and calm weather is somehow associated with earthquakes. Aristotle believed that such calm days were

caused by large amounts of
and then escaping force
during the day or
calm or storm

Earthquake
tions of the plane
producing an espec
brings on earthquakes
alignments, of course, bu
earthquakes. That notion
caverns puffed up with air. 1
Ju-Ju or Boo-Boo has been m
his people and so orders an earth
is also still with us today in some p

DANCE
OF
THE
CONTINENTS

Earth
the m
arou
sh
b

The Causes of Volcanoes

The causes of volcanoes also have their ju
history. The fiery outbursts of the volcanic n
by the Romans supposedly were the work of V
fire and metalworking. A volcano erupted whenev
up his forges to make weapons for Mars, the god of
mountains were called "vulcanoes," later changed to

One of the greatest geographers of ancient time
Greek scholar and traveler named Strabo, born around 6
His great book, *Geography*, was a huge task. He echoed a n
ber of Aristotle's ideas. Among them was the notion that volc
noes erupted for the same reason earthquakes occurred. Winds
blowing into and becoming trapped within Earth's many cavernous hollows caused a build up of great pressure. When the
pressure became strong enough, volcano "safety valves" were
triggered, a volcano erupted, and the pressure was eased.

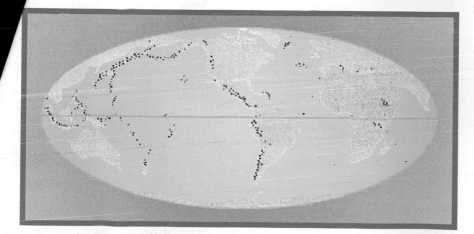

...quakes occur most often around the rim of the Pacific Ocean, along ...id-ocean ridge between the Americas and Europe and Africa, and ...nd the Mediterranean Sea. Major earthquakes near the surface are ...wn by squares and large round dots. Deep-focus earthquakes are shown ...triangles.

Compare this world map of active volcanoes with the world map showing the most active areas of earthquakes. Notice that most volcanoes also are located along the rim of the Pacific Ocean, which clearly indicates that quakes and volcanoes are related. Also compare these maps with the m... on page 57 that shows the world's major plates. Notice that the rim of... Pacific Plate, for example, marks the sites of both volcanoes and q... Where plates come in contact, expect action.

Earthquakes were felt right along with the volcanic eruption. This volcano safety valve theory was to linger on for nearly two thousand years.

Strabo said that the islands near Italy just northeast of Sicily were formed by volcanic action, or possibly by earthquakes. He correctly said that the lands of Lower Egypt, where fossils of sea animals were found, were once covered by an ocean. He also correctly described the ability of rivers to build up sprawling deposits of mud at their mouths, such as the vast delta land at the mouth of the Nile River. Although his idea that the land and seas change positions through time was correct, earthquakes were not the cause.

In the long search for causes of Earth's often frightening behavior, there have been two types of explanations. They have included natural causes, such as those offered by Strabo and Aristotle. And there have been explanations based on myth and superstition. Interestingly, it is often the fanciful explanations that get the most attention. It is also interesting that such explanations tend to hang on and survive, some into modern times. Because they do, they, too, have earned a place in the history of people's ideas about the forces that make Earth the active planet it is.

Volcanoes in Myth

Anyone who has been close to an erupting volcano will agree that it can be a terrifying event. Clouds of ash and hardened lava are exploded thousands of feet skyward. Firebombs are flung great distances from the volcano's crater. Lightning and thunderous roars add to the chaos as lava gushes down the volcano's slopes and the ground trembles. Since people of long ago had no way of knowing about the natural causes of a volcanic eruption, many let their imaginations invent causes. If the actual cause is not evident, then invent one.

Among the myths that have been woven to explain why a volcano erupts are those of the people of Central and South America—the Aztecs, Mayas, and Incas. A god or goddess becomes angered by the behavior of the people and threatens them with destruction in the form of a volcanic eruption. To soothe the angered god, the high priests offered human sacrifices by throwing children or young adults into the fiery crater. Although human sacrifices no longer are made, some cultures still do the next best thing. On the Pacific Island of Java, tribal leaders to this day throw live chickens into the crater of Mount Bromo once a year.

Also, to this day, the goddess Pele of the Hawaiian Islands is alive and well, at least in the imaginations of her believers. Whenever lava from an erupting volcano threatened to engulf a Hawaiian village, a member of the royal family would be called on to calm Pele's anger and so stop the lava flow. In August 1881, Mauna Loa threatened the city of Hilo. Princess Ruth Keelikolani rushed to Hilo and marched to the very edge of the advancing molten rock. She then began chanting in her ancient language, made offerings of silk scarves to Pele, and poured brandy onto the advancing lava. The next morning the lava flow stopped, and the village was saved. Who could deny that the ritual had not worked?

As recently as 1955 a lava flow threatened the Hawaiian village of Kapoho. People gathered at the advancing edge of the flow and began chanting. They also offered Pele food and tobacco. Once again, the lava flow stopped short of the village.

One of the world's liveliest regions of volcanoes is Iceland. And the most feared of its volcanoes is Hekla. Icelanders today caused by large amounts of air flowing into Earth's hollows and then escaping forcefully. The fact is that earthquakes occur during the day or night, or when it's cold or hot, or when it's calm or stormy.

Earthquakes have also been blamed on the changing positions of the planets. We read about "alignments of the planets" producing an especially strong gravitational tug on Earth that brings on earthquakes. From time to time there are planetary alignments, of course, but the alignments have nothing to do with earthquakes. That notion is every bit as false as Aristotle's vast caverns puffed up with air. The ancient belief that the great god Ju-Ju or Boo-Boo has been made angry by the misbehavior of his people and so orders an earthquake as a way of punishment is also still with us today in some parts of the world.

The Causes of Volcanoes

The causes of volcanoes also have their jumbled roots in ancient history. The fiery outbursts of the volcanic mountains observed by the Romans supposedly were the work of Vulcan, the god of fire and metalworking. A volcano erupted whenever Vulcan fired up his forges to make weapons for Mars, the god of war. So these mountains were called "vulcanoes," later changed to volcanoes.

One of the greatest geographers of ancient times was a Greek scholar and traveler named Strabo, born around 64 B.C. His great book, *Geography*, was a huge task. He echoed a number of Aristotle's ideas. Among them was the notion that volcanoes erupted for the same reason earthquakes occurred. Winds blowing into and becoming trapped within Earth's many cavernous hollows caused a build up of great pressure. When the pressure became strong enough, volcano "safety valves" were triggered, a volcano erupted, and the pressure was eased.

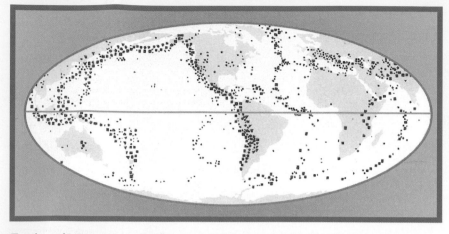

Earthquakes occur most often around the rim of the Pacific Ocean, along the mid-ocean ridge between the Americas and Europe and Africa, and around the Mediterranean Sea. Major earthquakes near the surface are shown by squares and large round dots. Deep-focus earthquakes are shown by triangles.

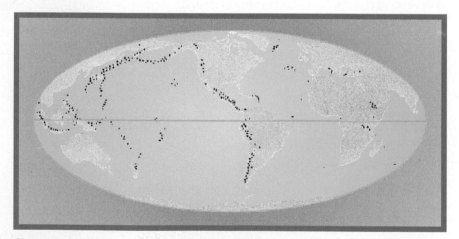

Compare this world map of active volcanoes with the world map showing the most active areas of earthquakes. Notice that most volcanoes also are located along the rim of the Pacific Ocean, which clearly indicates that quakes and volcanoes are related. Also compare these maps with the map on page 57 that shows the world's major plates. Notice that the rim of the Pacific Plate, for example, marks the sites of both volcanoes and quakes. Where plates come in contact, expect action.

caused by large amounts of air flowing into Earth's hollows and then escaping forcefully. The fact is that earthquakes occur during the day or night, or when it's cold or hot, or when it's calm or stormy.

Earthquakes have also been blamed on the changing positions of the planets. We read about "alignments of the planets" producing an especially strong gravitational tug on Earth that brings on earthquakes. From time to time there are planetary alignments, of course, but the alignments have nothing to do with earthquakes. That notion is every bit as false as Aristotle's vast caverns puffed up with air. The ancient belief that the great god Ju-Ju or Boo-Boo has been made angry by the misbehavior of his people and so orders an earthquake as a way of punishment is also still with us today in some parts of the world.

The Causes of Volcanoes

The causes of volcanoes also have their jumbled roots in ancient history. The fiery outbursts of the volcanic mountains observed by the Romans supposedly were the work of Vulcan, the god of fire and metalworking. A volcano erupted whenever Vulcan fired up his forges to make weapons for Mars, the god of war. So these mountains were called "vulcanoes," later changed to volcanoes.

One of the greatest geographers of ancient times was a Greek scholar and traveler named Strabo, born around 64 B.C. His great book, *Geography*, was a huge task. He echoed a number of Aristotle's ideas. Among them was the notion that volcanoes erupted for the same reason earthquakes occurred. Winds blowing into and becoming trapped within Earth's many cavernous hollows caused a build up of great pressure. When the pressure became strong enough, volcano "safety valves" were triggered, a volcano erupted, and the pressure was eased.

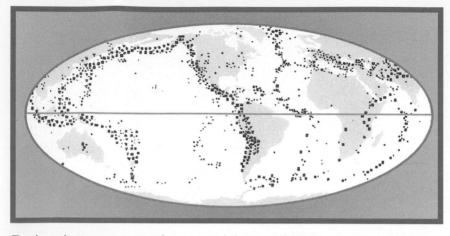

Earthquakes occur most often around the rim of the Pacific Ocean, along the mid-ocean ridge between the Americas and Europe and Africa, and around the Mediterranean Sea. Major earthquakes near the surface are shown by squares and large round dots. Deep-focus earthquakes are shown by triangles.

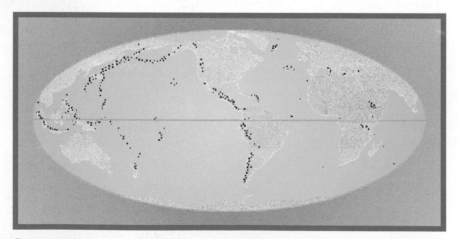

Compare this world map of active volcanoes with the world map showing the most active areas of earthquakes. Notice that most volcanoes also are located along the rim of the Pacific Ocean, which clearly indicates that quakes and volcanoes are related. Also compare these maps with the map on page 57 that shows the world's major plates. Notice that the rim of the Pacific Plate, for example, marks the sites of both volcanoes and quakes. Where plates come in contact, expect action.

Earthquakes were felt right along with the volcanic eruption. This volcano safety valve theory was to linger on for nearly two thousand years.

Strabo said that the islands near Italy just northeast of Sicily were formed by volcanic action, or possibly by earthquakes. He correctly said that the lands of Lower Egypt, where fossils of sea animals were found, were once covered by an ocean. He also correctly described the ability of rivers to build up sprawling deposits of mud at their mouths, such as the vast delta land at the mouth of the Nile River. Although his idea that the land and seas change positions through time was correct, earthquakes were not the cause.

In the long search for causes of Earth's often frightening behavior, there have been two types of explanations. They have included natural causes, such as those offered by Strabo and Aristotle. And there have been explanations based on myth and superstition. Interestingly, it is often the fanciful explanations that get the most attention. It is also interesting that such explanations tend to hang on and survive, some into modern times. Because they do, they, too, have earned a place in the history of people's ideas about the forces that make Earth the active planet it is.

Volcanoes in Myth

Anyone who has been close to an erupting volcano will agree that it can be a terrifying event. Clouds of ash and hardened lava are exploded thousands of feet skyward. Firebombs are flung great distances from the volcano's crater. Lightning and thunderous roars add to the chaos as lava gushes down the volcano's slopes and the ground trembles. Since people of long ago had no way of knowing about the natural causes of a volcanic eruption, many let their imaginations invent causes. If the actual cause is not evident, then invent one.

Among the myths that have been woven to explain why a volcano erupts are those of the people of Central and South America—the Aztecs, Mayas, and Incas. A god or goddess becomes angered by the behavior of the people and threatens them with destruction in the form of a volcanic eruption. To soothe the angered god, the high priests offered human sacrifices by throwing children or young adults into the fiery crater. Although human sacrifices no longer are made, some cultures still do the next best thing. On the Pacific Island of Java, tribal leaders to this day throw live chickens into the crater of Mount Bromo once a year.

Also, to this day, the goddess Pele of the Hawaiian Islands is alive and well, at least in the imaginations of her believers. Whenever lava from an erupting volcano threatened to engulf a Hawaiian village, a member of the royal family would be called on to calm Pele's anger and so stop the lava flow. In August 1881, Mauna Loa threatened the city of Hilo. Princess Ruth Keelikolani rushed to Hilo and marched to the very edge of the advancing molten rock. She then began chanting in her ancient language, made offerings of silk scarves to Pele, and poured brandy onto the advancing lava. The next morning the lava flow stopped, and the village was saved. Who could deny that the ritual had not worked?

As recently as 1955 a lava flow threatened the Hawaiian village of Kapoho. People gathered at the advancing edge of the flow and began chanting. They also offered Pele food and tobacco. Once again, the lava flow stopped short of the village.

One of the world's liveliest regions of volcanoes is Iceland. And the most feared of its volcanoes is Hekla. Icelanders today are all too knowledgeable about the natural forces that make Hekla and her sister mountains of fire erupt. But that was not so in earlier times. Icelanders long ago believed that the giant Surtur one day would destroy the world with his fire. About nine hundred years ago, Icelanders supposed that Hekla was the main entrance to hell. When it erupted, people claimed that for miles around they could hear loud wailing, mournful cries of lost souls, fearful howling, weeping, and the gnashing of teeth. Superstitions die hard.

Three

The Empty Years

The Middle Ages

What happened over the centuries after the ancient Greek philosopher–scientists? We might expect that there was a gradual growth of knowledge about the forces that have shaped and reshaped Earth's surface. But hardly anything new came about for a very long time. The years from about A.D. 450 to 1450 are known as the Middle Ages. They were dark times when scholars were all but forbidden to seek out new ideas about Earth below or the sky above. Christian and Islamic religious leaders preached that the only worthwhile knowledge was knowledge of God revealed in the holy books. If books written by the great scholars of earlier times did not praise God's work, then the books must

Illuminated manuscripts of the Middle Ages often show religious figures, but this one from the 1400s depicts what appear to be mountains. It is from On the Properties of Things, *by Bartholomaeus Anglicus.*

be burned. So one after another, the great libraries and centers of learning in the Western world were destroyed.

The only ideas permitted to survive openly were those approved by Christian and Islamic leaders. Aristotle's ideas were among them. For some reason, even though the Bible seemed to say that Earth was flat, the religious leaders accepted Aristotle's evidence that Earth was round. But they denied that Earth moved, either by turning on its axis or by circling the Sun. They taught that Earth was solidly fixed in space and was the center of all creation. Earth's land was divided into three continents—Europe, Asia, and Africa. All the rest of the planet was a

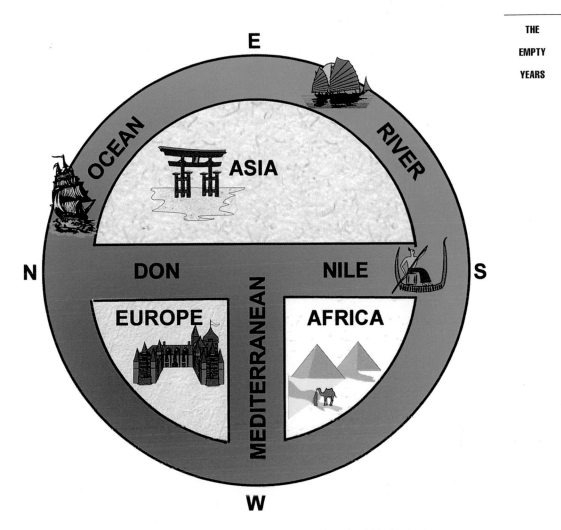

The Hereford Mappa Mundi (world map kept in Hereford Cathedral, England) is the most perfectly preserved world map of the Middle Ages. It is a T-O map, where the world is divided into three parts—the T within an O. The O is formed by the green band that represents the world ocean (the blue band in the diagram). The green T in the map (blue in the diagram) is formed by the rivers Don and Nile flowing into the Mediterranean Sea. The three known major land masses were Asia, Europe, and Africa. Notice that Jerusalem was located at the center of the world. East is at the top, and Christ at the Last Judgment crowns the map. The Garden of Eden was located at the top just within the world ocean boundary.

vast world ocean. Three great bodies of water separated the three land masses. They were the Don and Nile rivers and the Mediterranean Sea.

Mud, clay, and other sediments flushed out of rivers and built up as new delta land seemed to be one of the major causes of land change. Another was the erosion of mountains by rain and wind. In the year 1000, Aristotle's howling winds pent up in great caverns within the planet were still thought to be the causes of earthquakes, volcanoes, and mountain building. That belief was to last for centuries more. And so was Aristotle's explanation for the formation of rocks and minerals. He said that "dry exhalations" from deep inside Earth turned into rocks, while "moist exhalations" became metals. Exactly what Aristotle meant by "exhalations" is hard for us to understand in modern times.

People during the Middle Ages also supposed that faraway lands were inhabited by races of monsters. There were headless beings with eyes and mouths as parts of their chests. And the far reaches of the atmosphere, all the way to the Moon, were inhabited by angels and demons.

The world was pretty much as Strabo had left it.

Four

Digging for New Facts

We next move ahead to the 1600s and 1700s, 150 years after the close of the Middle Ages. Many of the old and erroneous opinions of Aristotle began to be questioned. In their place came careful observation of the land and experiments with Earth's materials. Those two approaches eventually led to the science of *geology*. Geologists examine Earth's composition and structure as revealed by its rocks and its history. But many of the old ideas rooted in religion and superstition did not die overnight.

Earth's Surface and Noah's Flood

In 1681 Thomas Burnet, a clergyman at England's Cambridge University, tried to explain Earth's mountains, valleys, and other surface features by following the account of the Flood in the Bible. He imagined that Earth in its youth was a smooth-skinned planet partly filled with water. After years of being heated by the Sun, Earth's crustal rock cracked and broke open. Enormous blocks of land plunged into the subterranean ocean. As they did, they sent gigantic waves surging overland, devastating all they touched. Mountains were thrust up, and deep valleys were gouged out. Great chunks of Earth's crustal rock were folded, tumbled, and twisted this way and that. In Burnet's thinking, the biblical Flood left Earth a shattered and chaotic globe.

Although scholars and clergymen alike attacked Burnet's theory, many went along with him, at least for part of the way. They agreed that the Flood most likely raised Earth's mountains. If not, they argued, how do we account for seashells entombed in mountain rock? Surely these are remains of creatures washed high by the Flood. Burnet and his followers were on the right track, but for the wrong reasons.

Earlier, around 560 B.C., the ancient Greek scholar Xenophanes had taken a special interest in *fossils*, which are the ancient remains of animals and plants and clues to past life. He had found fossils of fish far inland. Many people wondered how a fish fossil could end up on a mountain far from the ocean. Xenophanes said that in the dim past parts of the land had been covered by sea. Later the sea floor was thrust up as a mountain range, and so the fish fossils were left high and dry. During the Middle Ages, many people believed that fossils were the work of the devil when he tried unsuccessfully to create animals.

Just before Burnet's time, around 1500, the Italian Leonardo

Some 570 million years ago marked the heyday of trilobites, ocean-living relatives of crabs. They became the world's dominant animals and survived some 230 million years before becoming extinct. The fossil remains of some 10,000 species can be found high and dry, such as this group of prize specimens discovered in an abandoned Ohio cement quarry. Trilobites and other marine fossils are proof that sea floors rise and become dry land, once a notion hard to understand.

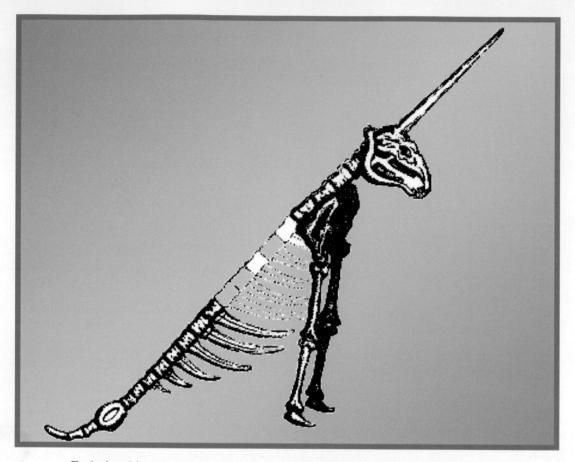

Early fossil hunters often didn't know much about the bones they uncovered. Here is one early attempt to assemble odds and ends of unrelated fossil bones to reconstruct the mythical beast called a unicorn, with a horn in the middle of its forehead. Earlier, in the Dark Ages, many people believed that fossils were the work of the devil when he tried unsuccessfully to construct animals.

da Vinci had also wondered about the fossils of sea animals he found high among the mountains of northern Italy. He explained that after the animals died and sank to the ocean bottom, they were covered over by mud and other sediments washed off the land. The sediments eventually turned into rock, and later the old sea floor was uplifted as land. Wearing away of the rock eventually exposed the fossils. Not only was Leonardo correct, but he was among the early thinkers to suppose that certain kinds of rock, over long periods of time, are formed from soft sediments. Exactly how, Leonardo could not say. Today we know that over thousands of years the sediments are compressed and heated by the great weight of thousands of feet of sediments pressing down from above. The combination of heat and pressure is what turns the soft sediments into hard rock.

The idea of a worldwide catastrophe caused by the biblical Flood struck many geologists around Burnet's time as a sound starting point to understand the surface features of the land. One was the German geologist Abraham Gottlob Werner, born in 1750. On studying mine shafts in his native Saxony, Werner saw that the exposed dirt walls were arranged in layers, or *strata*, as in a layer cake. Werner was not a traveler, so he assumed that the same kinds of strata probably were found the world over. He also assumed, incorrectly, that those rock layers had formed one atop another in just the same order in China or Texas as in his native Saxony. His problem was to explain how the strata were formed.

Werner further supposed that there had been many great floods of biblical size in Earth's history. Each time, a vast ocean covered the land and then dried up. This happened over and over again, each time a new layer of sediments being deposited. He imagined Earth being composed of layers like those of an onion, each layer having formed during a given flood period. Because Werner looked to a world ocean to explain rock strata, his followers were called Neptunists, from the Roman god of the sea, Neptune.

If Werner were correct, then all the rock layers he studied

should have been sedimentary rocks. But he was confused whenever he came across a certain kind of rock that his sediments theory could not account for. Called *basalt*, it was dark gray to black and had very fine grains. The French geologist Nicolas Desmarest, born in 1725, became especially interested in these basalt rocks. He found many of them strewn about the extinct volcanoes of the Auvergne Mountains in central France. He felt that the rocks were not formed from ancient ocean sediments, but from the old volcanoes. Today we include basalt rocks in a larger group called *igneous* rocks, from the Latin word for fire.

Desmarest concluded that in the distant past the ancient active volcanoes had emitted huge lava flows that forced their way up through cracks in the surface sedimentary rock. On hardening, the lava became basalt. Desmarest's followers, who placed great importance on the role of volcanoes to account for many of Earth's land formations, were called Vulcanists, after the Roman god of fire, Vulcan.

No Beginning, No End

If any one person can be singled out as "the father of geology" it is the Scottish physician–farmer–chemist–geologist James Hutton. Around 1790 he published a book titled *Theory of the Earth*. He wrote that Earth has "no vestige of a beginning, no prospect of an end." Our planet was untold millions of years old,

A. G. Werner supposed that a vast ocean covered and then uncovered the land from one period to the next. Each time the ocean withdrew, he supposed that a new layer of sediments was laid down. Because he and his followers looked to a world ocean as the cause of sedimentary rock layering, they were called Neptunists, after the Roman god of the sea, Neptune. This water fountain statue of Neptune decorates the grounds of Russia's Peterhoff palace and pays honor to the king of the sea.

he said. He imagined a world of eternal change brought about by natural forces. Great upheavals within Earth produced changes on the surface, and rain, wind, and rivers continually sculpted Earth's face. He said that the forces at work today were the same forces that have been at work over those millions of

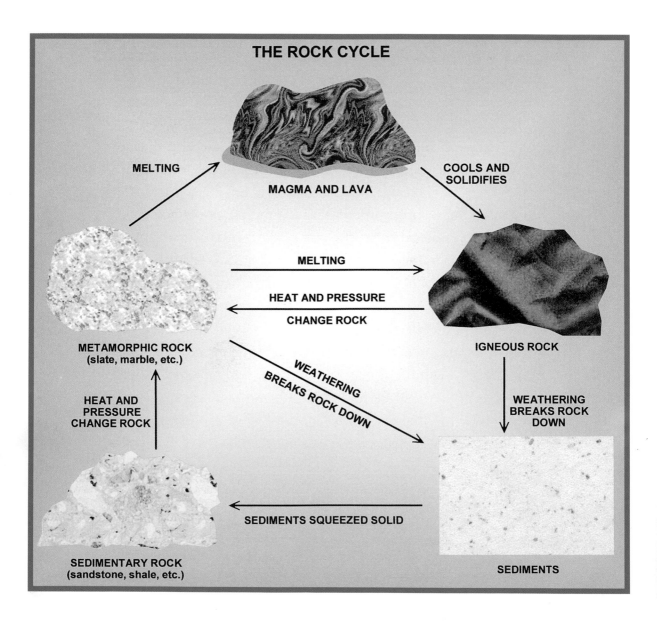

THE ROCK CYCLE

MELTING

MAGMA AND LAVA

COOLS AND SOLIDIFIES

MELTING

HEAT AND PRESSURE CHANGE ROCK

METAMORPHIC ROCK
(slate, marble, etc.)

IGNEOUS ROCK

HEAT AND PRESSURE CHANGE ROCK

WEATHERING BREAKS ROCK DOWN

WEATHERING BREAKS ROCK DOWN

SEDIMENTS SQUEEZED SOLID

SEDIMENTARY ROCK
(sandstone, shale, etc.)

SEDIMENTS

years. Therefore, the present is a key to the past. That idea came to be known as the principle of *uniformitarianism*, which every student of geology the world over is taught.

Hutton correctly maintained that great heat within Earth melted solid material, which eventually hardened and reformed as igneous rocks. He also said that this heat long ago caused many mountains to rise up higher than the seas. Sedimentary rocks, he said, are formed over many centuries as rain and rivers wash sediments into the sea. Slowly those sediments are packed and heated under great pressure to become layered beds of rock. He also held that heat and pressure deep within Earth's crustal rock change the sedimentary rock limestone into marble. In short, Hutton's "Plutonism" (named after the Greek god of the underworld, Pluto) explained all rocks found the world over.

But Hutton had critics. Most were fundamentalist Christians, who believed every word of the Bible to be true, both as history and as science. They could not accept natural forces being responsible for geologic change. All was the work of God, they believed. Furthermore, they would not accept an age of hundreds of millions of years for Earth. Bible scholars said that Earth had to be much younger, only six thousand years old. To this day, Christian fundamentalists believe that Earth is not older than ten thousand years. Geological dating, accepted by scientists the world over, puts Earth's age at 4.6 billion years.

Hutton had other critics, as well. They included the Neptunists, who denied that basalt could be formed from the outpourings of lava from volcanoes. They also argued that when limestone was heated, it turned to quicklime, not marble. On these points a Scottish friend of Hutton's, Sir James Hall, came to the rescue. Hall packed limestone into a gun barrel and heated it under pressure. The limestone was changed to marble, not quicklime. And Hall showed that lava, when melted and cooled, could be turned into either a smooth, black, glasslike substance called obsidian, or into basaltlike rock. Slow or rapid cooling determined what would happen to the lava. Cooled slowly, it became basalt. Cooled quickly, it became obsidian. Here was proof that Hutton was correct.

Hutton's work did much to pave the way for one of the greatest geologists of all time, Sir Charles Lyell. In 1830 Lyell published his famous *Principles of Geology*. In these three great books he reviewed all the important studies in geology that had been made up to his time. He rejected the biblical young age for Earth and instead said that Earth was at least 240 million years old. He strongly attacked the old myths of geology, including the Flood geology so popular in the 1700s and into the 1800s. In fact, he ridiculed the idea that biblical-type floods could move mountains. He wrote that "the land has never in a single instance gone down suddenly for several hundred feet at once.... Great but slow oscillations brought dry land several thousand feet below sea level and raised it thousands of feet above."

Change Is Eternal

Lyell also held to the idea of eternal change. Even today, he said, the forces that build mountains, warp and crack the crust, produce strata, and cause the formation of fossils are still going on. But geologic time is long compared with human time. It is so long that in our brief life spans we see little of the continual change about us. Sometimes that change comes in spurts, other times in slow motion. With the publication of Lyell's three great books, buttressed by the work of Hutton, geology came of age in the early 1830s.

If we had to choose only one of Lyell's ideas as the one that

The English geologist Sir Charles Lyell rejected the biblical age for Earth and instead said that Earth was at least 240 million years old. He further denied the popular belief of the 1700s and early 1800s that biblical-type floods could move mountains. Lyell, the greatest geologist of his time, lived from 1797 to 1875.

brought about a major revolution in geological thinking, what would it be? It would have to be the forces that build mountains and that warp and crack Earth's crust. Why? In the early 1900s a German scientist unleashed an idea so powerful that it set the heads of geologists the world over spinning. Many simply laughed at the idea. Others just shook their heads in utter disbelief. Still others shrugged and called the unknown scientist a crackpot. And well they might have. The scientist wasn't even a geologist. He was an astronomer turned meteorologist and an Arctic explorer. His name was Alfred Wegener. His bold new idea was to haunt the world of geology for fifty years before his critics stopped laughing and could no longer shrug it off.

Charles Lyell held that the forces that long ago shaped Earth's crust are still going on today. Among such changes are mountain building and the endless laying down of sediments that harden as rock strata. This layering of sedimentary rock can be seen in the Kodachrome Basin of southern Utah.

Five

Wegener On Trial

The South America–Africa Connection

A glance at a map shows something rather interesting about South America and Africa. If you could slide the two land masses together, South America would fit snugly up against Africa. Is it possible that the two land masses actually were joined long ago as one? The fit is so obvious that several people, as early as the 1600s, asked that same question.

One was Francis Bacon, in 1620. The fit puzzled him. So did the larger fit of the Western Hemisphere to both Europe and Africa. Bacon said that maybe those large land masses all were joined in the distant past. Around 1858 the Italian Antonio Snider-Pellegrini said that perhaps *all* the continents were joined in the past. The similarity of 300-million-year-old fossil plants found in Europe and North America gave him the idea. However, he turned to the Bible and Noah's Flood for an explanation of the split up of that huge land mass.

It wasn't until the 1900s that the first scientific suggestions that the continents were once joined were made. Among the first was the Austrian geologist Eduard Suess. He pointed to the close agreement among rock formations in the lands of the Southern Hemisphere. The formations were so much alike, he said, that all of the Southern Hemisphere land masses must once have formed a single supercontinent, which he called Gondwanaland. The name comes from Gondwana, a geological province in east-central India. Then, in 1908, the German scientist Alfred L. Wegener came up with one of the first explanations of the forces that might account for the continents breaking apart and moving about.

Born in Berlin in 1880, Wegener eventually was to become an astronomer. He also was fascinated by Greenland and longed to explore that bleak world of ice. And he became especially interested in the then-new science of meteorology. In fact, he abandoned astronomy for meteorology.

We are not sure how long Wegener thought about the possibility of the continents moving about like gigantic stone rafts floating in a sea of molten rock. We do know that he first officially announced his theory of *continental drift* in January 1912 in a lecture before the German Geological Association. It must have taken a lot of courage to deliver that speech, for the idea that the continents might move about went against just about everything geologists then believed about Earth. They thought that Earth had been cooling down, solidifying, and shrinking since its early years as a sphere of molten rock. During its molten stage, they said, the heavier elements, such as iron and nickel, sank into the central core region. The lighter elements, such as silicon and aluminum, meanwhile floated up toward the surface, where they later formed the rigid crust we walk on today.

In the early 1900s the German scientist Alfred Wegener said that the continents move about like gigantic stone rafts floating in a sea of molten rock. Most geologists of the time scoffed at the idea, but today the idea of continental drift cannot be doubted. Wegener lived from 1880 to 1930.

As the "Big Apple" Cooled and Wrinkled

Most geologists felt quite comfortable with the cooling Earth model. It seemed to account nicely for many of Earth's surface features. Mountain ranges were wrinkled up as the planet's surface shrunk, something like the wrinkles formed on the skin of a baked apple. At the same time some land areas were driven down and became ocean basins. Others were left sticking up out of the chaos as continents. That could account for the land sometimes being high and dry but at other times sinking and becoming the ocean floor. Although geologists of the time agreed that the continents could move up and down, no one would admit that they could move sideways.

By Wegener's time the same kinds of plant and animal fossils had been found on continents far removed from each other. If the continents didn't slip and slide sideways, then how could those identical plants and animals have managed to populate lands so far away from each other? The answer was something called land bridges. These were sections of submerged land that from time to time became exposed as sea level lowered. Just such a land bridge allowed the migration of people and animals from Siberia to Alaska some ten thousand and more years ago. The sedimentary layers of rock that so interested Werner were evidence of the repeated rise and fall of oceans through geologic time.

Wegener was well aware of the contracting Earth model and how it was supposed to account for the geology of Earth's crustal rock. But he saw a number of features that the model couldn't explain. One was the neat fit of South America and Africa. Another was the location of the world's great mountain ranges. If the ranges actually had been wrinkled up by Earth's shrinking, like the skin of a baked apple, then why weren't the mountain chains arranged more or less evenly all over the globe? Instead they were found in narrow curving belts. Wegener further told his audience that his model of continental drift could explain why certain rock formations had stretched unbroken

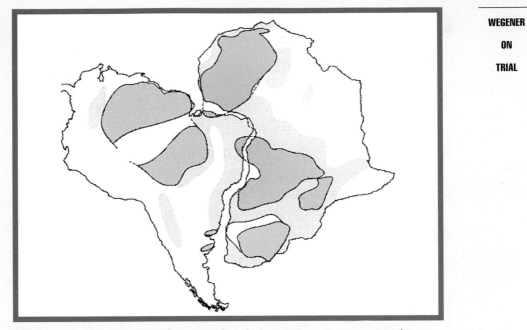

Geological evidence that Africa and South America were once joined came with the discovery that certain rock formations spanned the present boundaries of the two continents before they drifted apart. Such rock formations that once stretched unbroken across the two land masses are shown in light blue and green.

across South America to Africa before the continents split apart.

Wegener's audience listened politely as he explained how his model worked. He said that some 200 million years ago there was only one supercontinent, which he called Pangaea (from the Greek word meaning "all land"). Wegener asked his listeners why the continents could not move sideways, since they were able to move up and down. Why should the sea of putty-like rock in which the continents floated permit one movement but not the other?

Those who attended his lecture and then later read his book

clearly were interested in continental drift, if not entirely won over by the idea. By 1921 Wegener was pleased to be able to say that he knew of no major geologist who opposed his revolutionary theory. But at that time his theory was not widely known outside Germany. But over the next several years it gained attention, and criticism, from geologists in England and the United States. And the criticism was not what you would call polite.

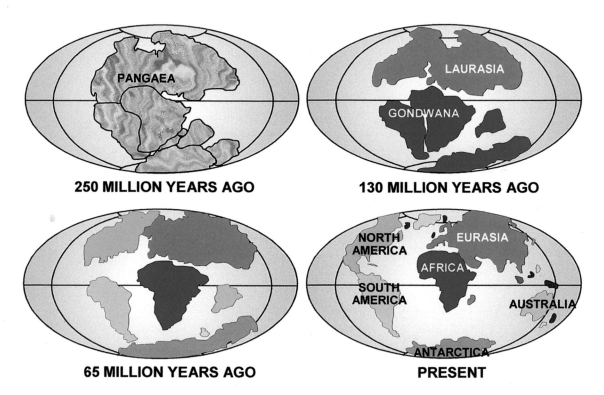

250 MILLION YEARS AGO   **130 MILLION YEARS AGO**

65 MILLION YEARS AGO   **PRESENT**

Wegener imagined that some 250 million years ago an enormous supercontinent called Pangaea made up Earth's surface. Over time, Pangaea was broken apart by plate movements into two lesser continents, one in the north called Laurasia, and the other in the south called Gondwana. By about 65 million years ago, the continents had further broken apart. Further drifting positioned them in the locations where we see them today.

Humiliation and Defeat

One of Wegener's critics was the famous English geologist Sir Harold Jeffreys. He attacked Wegener's theory where it was weakest. What forces within the planet could possibly cause the continents to drift, he demanded to know? Earth's crustal rock and the deeper rock beneath were too rigid to allow such large scale movement, he said. Jeffreys was highly respected, and his opinions carried much weight. Wegener did not have a convincing answer. He mentioned something about "tides in Earth's crustal rock." As evidence, he then pointed out that India was forcefully moving northward and wrinkling up the great Alps and Himalayan mountains. Jeffreys said that was impossible. Earth's crustal rock is far too strong to be moved even slightly that way. If Earth allowed that kind of motion, Jeffreys argued, then the huge mountain ranges would sink under their own weight. Wegener had no answer and had to abandon his "tidal forces" idea.

The most hostile criticism came in 1928 from geologists attending a meeting of the American Association of Petroleum Geologists. Like a pack of vicious dogs, one after another tore into every scrap of evidence Wegener had offered. After demolishing his theory, some of his critics then attacked Wegener himself by saying that he wasn't even a geologist. What business did he have advancing his impossible theory before the world's leading experts in geology and Earth history? He was called an outsider. By the time they had finished with him, Wegener's theory, and Wegener the scientist, had been crushed. He had neither the energy nor the money to mount a vigorous defense. Wegener died two years after the 1928 meeting. He was on an expedition to Greenland, became lost, and apparently froze to death.

Six

The Continents Do Dance

New Evidence and Triumph

As is so often the case in all branches of science, geologists had to be overwhelmed with evidence before they would accept the theory of continental drift. That evidence was slow in coming, but come it did. By the 1960s there wasn't a geologist alive who didn't enthusiastically accept continental drift as a major geological fact of life. American geologists were the last to hold out as the theory's opponents. It wasn't until well into the 1960s that they finally came around to accepting the abundance of evidence proving that the continents *do* dance.

The ocean floor spreads as new floor material wells up from below and spills out of the great mid-ocean ridge called the Mid-Atlantic Ridge. In places, upwelling rock pokes mountain peaks above sea level. Such peaks in the North Atlantic include Iceland, and in the South Atlantic, Tristan da Cunha. The Mid-Atlantic Ridge did not exist 165 million years ago, and neither did the Atlantic Ocean. All the land masses now separated by the Atlantic were joined. You could have walked from New York to Paris. Today North America and Africa are creeping away from each other.

Sea Floor Spreading

A new view of the deep ocean floor—not of the land—won the day for moving continents. By the late 1950s scientists studying the deep sea floor had shown that many undersea mountain ridges formed a globe-circling chain some 43,000 miles (69,000 kilometers) long. Hundreds of miles wide in some places, the chain snaked its way around the planet and sometimes poked mountain peaks above sea level. Such peaks in the North Atlantic include those of Iceland, and in the South Atlantic, Tristan da Cunha. That ridge splits the Atlantic Ocean down the middle and is named the Mid-Atlantic Ridge. It did not exist 165 million years ago, and neither did the Atlantic Ocean. All the land masses now separated by the Atlantic were joined. You could have walked from New York to Paris.

As we move east or west from the mid-ocean ridge toward Europe or the United States, the ocean floor sediments grow deeper. We might think that those sediments of clay, mud, and

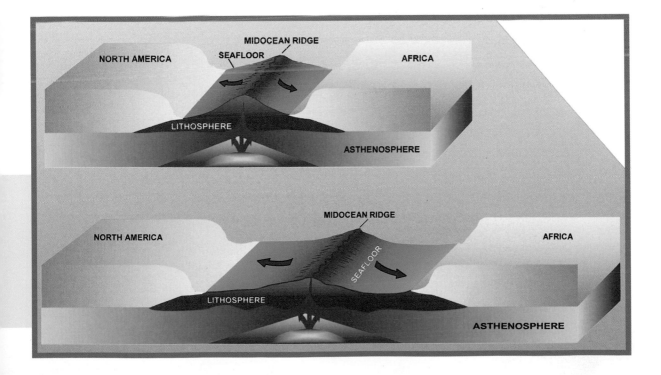

other materials washed off the continents and out onto the ocean floor would be spread out pretty evenly. The question oceanographers faced was why the sediments were shallow near the mid-ocean ridge but deep near the edge of the continents? Another thing they wondered about was that the rock floor near the mid-ocean ridge is younger than sections of floor farther away from the ridge.

The answer to both puzzles became clear when oceano-graphers discovered a *rift valley* all along the ridge top of the mountain chain. From time to time molten rock from deep within the planet wells up through the rift valley. As it does, it keeps spreading the rift valley wider. The molten rock then cools and forms new sea floor rock along the mid-ocean ridge.

With this and other information, around 1960 the Princeton University geologist Harry H. Hess proposed that the ocean floor moves and has been spreading apart over millions of years. Almost overnight his idea that the sea floor is gradually spreading outward on both sides of the mid-ocean ridge could not be denied. We now know that the sea floor of the Atlantic is widening by almost an inch (about two centimeters) a year. So the continental partners, Europe and the United States, are slowly dancing away from each other. Similar sea floor movement is now known to be taking place over the entire globe.

The discovery of sea floor spreading answered a lot of questions. For one, it explained why the ocean bottom rock near the mid-ocean ridge was younger than rock nearer the continents. Because it was younger, it had had less time to collect sediments washed off the land. So the mystery of shallow sediments along the ridge also was solved. Sea floor spreading showed that Jeffreys had been wrong and that Wegener had been right. Forces deep within the planet could and did permit the continents to move about, and without allowing the great mountain ranges to sink into oblivion.

Mountains making up the undersea ridge system are the most noticeable sea floor feature. As mountain ranges on a continent tower high above the land below, the mid-ocean ridges tower above the deep-ocean basins. But there the comparison ends. Continental mountain chains, such as the Alps or Himalayas, are formed mainly of sedimentary rock types that have been crushed and crumpled together. The mid-ocean ridges are made of enormously thick layers of the volcanic rock basalt. Over the past twenty years or so scientists have found dramatic evidence of the geologic inferno beneath the worldwide network of mid-ocean ridges.

Undersea Geysers

A vast store of molten rock, called *magma*, lies less than a half mile (less than a kilometer) beneath the base of the mid-ocean ridge system. Sea water seeps down through cracks in the ridge's rock, is heated, and is then forced back up to the sea floor through vents. The ejected hot water is rich in salts and other minerals. The outpourings of water are called *smokers*, although they have nothing to do with smoke. They actually are undersea geysers. Perhaps Aristotle would not be surprised and would say that he had told us about "exhalations" from Earth's interior.

There are white smokers and there are black smokers. The different kinds of minerals a smoker has dissolved and carries provides its color. Smokers that have sulfur-bearing minerals such as iron, zinc, lead, and copper are dark gray to black. Other smokers lacking these minerals are white. The water ejected from a smoker may be as hot as 625°F (330°C), evidence of the tremendous heat beneath Earth's crustal rock.

It is now thought that the entire volume of ocean water is cycled through the plumbing system beneath the sea floor every three million years or so. The chemical exchange between the ocean water and the hot rock below strongly affects the chemistry of the oceans. Long after rich mineral deposits build up as part of the mid-ocean ridges, sea floor spreading gradually carries the deposits to the continents. Today we mine valuable metal-bearing minerals, once part of the sea floor and later raised as land, in Arizona, Utah, Chile, and on the Mediterranean island of Cyprus.

Earth's Crust and Mantle

How do we know what lies beneath Earth's thin shell of crustal rock, which is only some 20 miles (32 kilometers) deep? The deepest mines go down only about 2.5 miles (4 kilometers). The world's deepest well is in a remote northern outpost of Russia, and it is a little less than 8 miles (13 kilometers) deep. Even volcanic eruptions bring material to the surface from a depth of only about 100 miles (160 kilometers)—a small fraction of the great distance to Earth's center.

Around the year 1700 the English scientist Edmond Halley thought that Earth's interior was made up of onionlike layers of hot gases. He further imagined that the glowing gases lit up inner Earth and from time to time escaped from the North Pole and caused displays of northern lights.

What we know about Earth's interior today does not come from any of the deep wells or mines. It comes from earthquake

Like clouds of black smoke billowing up out of the sea floor, deep ocean geysers bubble up hot mineral-rich water. This artist's view of one such geyser shows the abundance of plant and animal life associated with these cracks in the ocean floor called hydrothermal vents. When a vent becomes blocked, its community of plants and animals perishes.

waves. There's no shortage of earthquake waves since more than a million quakes send shivers through the planet each year. Since the early 1900s geologists have used earthquake waves to learn about Earth's deep interior. Over the years they have come to know the kinds of rock materials that make up the planet and how they are arranged from the surface down into Earth's center. They also have been able to map the planet's deep rock zones. Just as light rays are bent, or refracted, when they pass through a glass prism or through water, earthquake waves are also bent when they pass through layers of different kinds of rock. By studying the travel paths and speeds of earthquake waves, geologists have been able to learn about the lightweight granite-like rock that makes up the continents. They also have learned about the heavyweight basalt rock of the ocean floor, the much heavier rock materials that make up Earth's deeper region, and finally the mix of iron and nickel of the planet's core regions.

From the crust down to a depth of about 40 miles (65 kilometers) is a region of rigid and strong rock called the *lithosphere*. The word "lithos" comes from the Greek and means "stone." The rock layer beneath the lithosphere and extending to a depth of about 450 miles (725 kilometers) is called the *asthenosphere*. Its name comes from the Greek and means "without strength." The bottom of the asthenosphere marks the depth limit for earthquakes. Because rock of the asthenosphere is under tremendous pressure from the great weight of rock above, the material of the asthenosphere moves about like putty. It is just such movement within the asthenosphere that helps explain the moving about of continents and spreading of the sea floor.

The asthenosphere marks the upper region of Earth's interior, called the *mantle*. The mantle rock extends to a depth of about 1,800 miles (2,900 kilometers). The planet's center is an

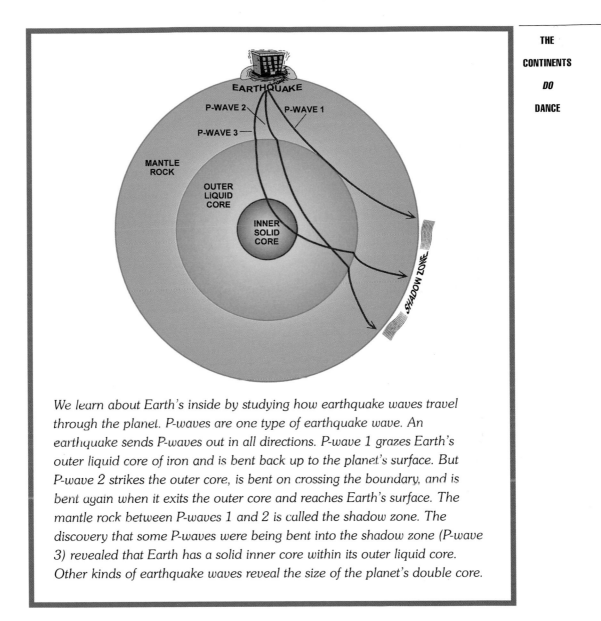

We learn about Earth's inside by studying how earthquake waves travel through the planet. P-waves are one type of earthquake wave. An earthquake sends P-waves out in all directions. P-wave 1 grazes Earth's outer liquid core of iron and is bent back up to the planet's surface. But P-wave 2 strikes the outer core, is bent on crossing the boundary, and is bent again when it exits the outer core and reaches Earth's surface. The mantle rock between P-waves 1 and 2 is called the shadow zone. The discovery that some P-waves were being bent into the shadow zone (P-wave 3) revealed that Earth has a solid inner core within its outer liquid core. Other kinds of earthquake waves reveal the size of the planet's double core.

extremely hot core of iron and nickel, a ball within a ball. The outer liquid core reaches down about 3,100 miles (5,000 kilometers). The center of the solid inner core is 4,000 miles (6,450 kilometers) beneath the surface.

Giant Rafts of Stone

By the 1960s a flood of new knowledge about Earth's interior had enlarged Wegener's theory of continental drift into a much more complete theory called *plate tectonics*. It began to look as if Earth's lithosphere were made up of a number of giant *plates*, or rafts of solid rock. It also looked as if those rafts moved about on the subterranean sea of puttylike rock, the asthenosphere. So tectonics, from the Greek words "to build," emphasizes the idea of plate movement. Today, the new science of plate tectonics cannot be denied.

The plates are like giant pieces of a jigsaw puzzle, but a puzzle that won't sit still. For millions of years the puzzle has been changing as the plates creep along. As they collide, ride up over or dive beneath one another, they reshape the continents and ocean basins alike. Whenever two plates scrape past each other, their edges stretch and snap, causing earthquakes that shatter cities. Plate interactions fire up volcanoes, raise the sea floor as dry land, and wrinkle up mighty mountain ranges. They also bend the sea floor into monstrous deep-sea trenches large enough to swallow up Mount Everest with room to spare.

The planet's crustal plates come in two general sizes. There are six large ones and about a dozen smaller ones. All move about slowly, on the average about as fast as your fingernails grow. While some move less than half an inch (1 centimeter) a year, others speed along at 3.5 inches (9 centimeters) a year. Over long periods of several hundred million years, plate movements have broken up giant continents into smaller ones and smashed other continents together. All such activity is created within the top 125 miles (200 kilometers) of Earth's restless skin.

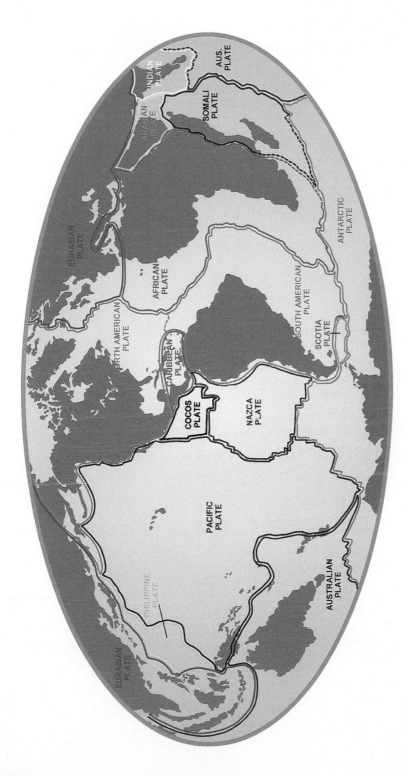

Cracked into a dozen or more plates, the planet's lithosphere is a complex of stone rafts slipping over a sea of hot plastic rock. Wherever plate edges grind against one another, there are earthquakes and volcanoes (see diagrams on page 20). In places, plates pull apart from each other and allow molten rock from beneath to well up and spread over the sea floor. One place this occurs is all along the 43,000-mile-long (69,000-kilometer-long) Mid-Atlantic

Ridge. The ridge traces a wiggly line from the Arctic down between North America and Europe and between South America and Africa to the Antarctic Plate. Another place where plates are pulling apart is the East African Rift Valley, marked by the eastern edge of the African Plate and the western edge of the Somali Plate. Still another place is western California where the Pacific Plate and the North American Plate are splitting apart.

Seven

How Plates Move

Plates Move in Three Ways

There are three types of plate movement. Each produces its own effects, whether on land or on the sea floor. One way is for two neighboring plates to pull away from each other. This is just what is happening to the South American Plate and the African Plate. As the South American Plate creeps westward, the African Plate creeps eastward. If two plates are pulling apart from each other, the cause can be that something is *pushing* them apart. That cause is well known. It is sea-floor spreading along the Mid-Atlantic Ridge.

If a continent or other large land mass happens to be sitting on top of such a ridge of upwelling magma, the land mass is split

*In several parts of the world magma upwelling beneath a large land mass breaks the land mass in two (**1**). A ridge of new rock crust then begins to form. As more magma keeps welling up, the ridge spreads outward on both sides (**2**). The spreading action pushes the split land mass pieces farther and farther apart. Today that is just what is happening to Iceland.*

apart. Today that is just what is happening to Iceland and, on a much larger scale, to the eastern third of Africa. Two branches of a several-thousand-mile-long rift valley run almost the length of the African continent. In the past, the Gulf of California was formed by the Pacific and North American plates pulling apart from each other. The Red Sea was formed by the Arabian and African plates pulling apart.

As two plates are pushed apart by volcanic outpourings along a rift valley, what happens to their opposite edges? We know by keeping track of the South American and Nazca plates. The two plates are colliding. While the South American Plate is being pushed westward by sea floor spreading along the Mid-Atlantic Ridge, the Nazca Plate is being pushed eastward by sea floor spreading caused by a second major rift valley called the East Pacific Rise.

When two plates collide there are bound to be fireworks. At ground level there are lots of volcanoes, lots of earthquakes, and lots of mountain building along the common edge where the two plates meet. That is how the mighty Andes mountain range, which runs the length of western South America, was shoved skyward.

But something even more interesting happens deep down at the common edge of the colliding plates. The westward edge of the lightweight continental rock of the South American Plate is riding up over the eastward edge of the heavyweight basaltic

When two plates collide, the lighter plate of less dense rock (the South American plate here) rides up over the heavier plate of denser rock (the Nazca plate). The forward part of the heavier plate is forced down into what geologists call a subduction zone. Rock forming the lower edge of the subducted plate is melted by the hot asthenosphere magma. The molten rock then rises, forces its way up through the lithosphere and plate above, and pours out as a volcano. South America's Andes Mountains were formed in this way, and the activity continues to this day.

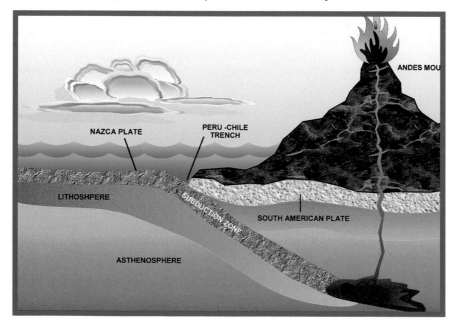

California's 750-mile-long (1,205 kilometers) scar in its rock crust is called the San Andreas Fault. It marks the line along which the North American and Pacific Plates rub against each other in opposite directions. But instead of sliding, the two plates tend to stick together and stretch. Every so often the rock gets stretched to its limit and snaps. At such times earthquakes shake the area. During the 1906 San Francisco earthquake parts of California along the San Andreas Fault lurched northward by 20 feet (6 meters).

rock of the Nazca Plate. This forces the leading edge of the Nazca Plate down into the asthenosphere of the upper mantle into what geologists call a *subduction zone*. As the plate edge is forced down, it pulls part of the ocean floor crust down with it. The result is the formation of a deep oceanic trench—the Peru-

Eighteen Major Earthquakes and Their Death Tolls

Year	Place	Deaths
1906	San Francisco	3,000
1908	Messina, Italy	110,000
1920	Gansu and Shanxi, China	200,000
1923	Tokyo, Japan	142,810
1927	Qinghai, China	200,000
1935	Quetta, Pakistan	30,000
1939	Erzincan, Turkey	32,700
1948	Fukui, Japan	5,390
1960	Southern Chile	5,700
1970	Northern Peru	67,000
1976	Mindanao, Philippines	8,000
1976	Guatemala	22,780
1976	Tangshan, China	655,000
1985	Michoacán, Mexico	9,500
1988	Armenia	25,000
1990	Western Iran	50,000
1993	Latur, India	9,750
1995	Kobe, Japan	5,200

Chile Trench—along the western shores of South America. As the forward edge of the Nazca Plate is bent and pushed down into the mantle, it is heated so much that it melts. This newly melted rock then forces its way up through the South American Plate. From time to time the fluid rock boils up to the surface and spills over the surrounding land as volcanic outpourings.

How else do plates move? Two plates can scrape and grind past each other at a common boundary that geologists call a *transform fault*. The zone of contact may run for a few to hundreds of miles, depending on the size of the plates. California's famous San Andreas Fault is an example. Some 750 miles (1,205 kilometers) long, it marks the common border of the North American and Pacific Plates. Year after year, the two plates rub in opposite directions and try to slide past each other along the fault line. But instead of sliding gently, friction causes the plate edges to stick together and stretch rather than slide freely. Then one day one of the rock faces reaches its limit of stretching. When it does, the plates snap along the fault line, and dishes begin to fall off shelves as houses crumble. During the 1906 San Francisco earthquake parts of California along the San Andreas Fault lurched northward by 20 feet (6 meters).

What Causes Plates to Move?

Measuring and agreeing about what Earth's plates are doing is one thing. Explaining the forces that make them move is quite another. One theory says that the plates move because they are being pulled along by the action of their forward edge diving down into the mantle. At the same time, along their opposite trailing edge thousands of miles away, the plates are being pushed apart and shoved along by sea floor spreading. If that is actually what happens, then we must ask what is going on deep in the mantle to cause sea-floor spreading. There is about as much disagreement among geologists as there is agreement.

The answer many geologists favor today is the *convection cell* theory. The rift valleys of the Pacific and Atlantic oceans lie along rising currents of magma. Heat deep within the mantle produces a conveyor belt motion of neighboring convection cells of molten rock. Where two neighboring cells turning in opposite directions meet, they push molten rock upward and out onto the ocean floor. This action creates the rift valley and keeps feeding it with new outpourings of magma. It is that action of the

Great convection cells of fluid rock are kept turning as hot magma wells up from the mantle. As the rising molten rock is pushed up, it forms a rift valley in the sea floor at plate edges. The continued addition of new rock outpourings causes the sea floor to spread outward from the rift valley. This is just what is happening today to the Atlantic Ocean floor due to outpourings from the mid-ocean ridge.

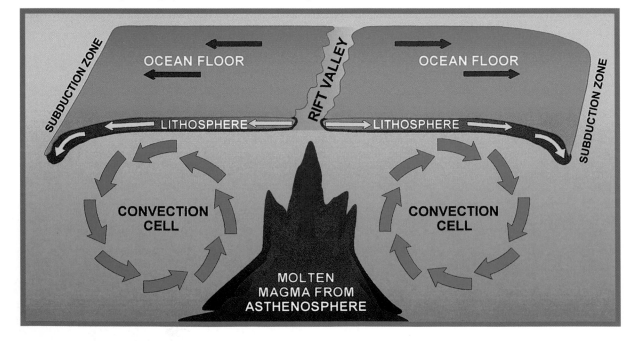

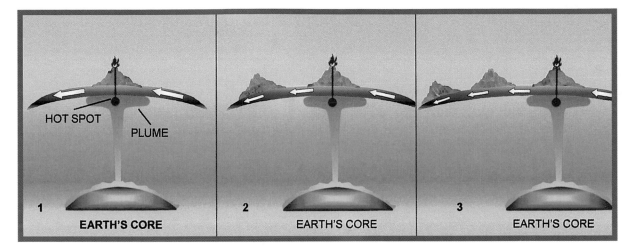

*Hot magma deep within Earth is thought to rise in columns called plumes. Plumes create hot spots beneath a plate. As a plate glides across a hot spot (**1**) magma wells up through the plate and builds a volcano. As the plate keeps moving, new volcanoes (**2** and **3**) are formed. In this way, the long chain of volcanic islands and seamounts stretching several thousand miles northwest from the Hawaiian Islands were formed by the Pacific Plate moving across a hot spot.*

convection cells that seems to drive seafloor spreading and keep plates of the lithosphere moving. But there may be another driving force in the upper mantle rock.

Many geologists look to the *plume* theory. They think that gigantic balloons of especially hot rock in the lower mantle rise into the upper mantle as *hot spots*. Over the past ten million years some 120 hot spots probably have been active, but only thirty or more seem to be active now.

Today hot spots boil away beneath Iceland in the North Atlantic and beneath Yellowstone National Park in Wyoming, for instance. Ancient hot spots probably are marked by enormous piles of old lavas, called flood basalts. They include the Columbia River Plateau in Idaho and Washington, the Deccan Plateau

in central India, and the Siberian Plateau within the Arctic Circle.
When the Deccan Plateau formed about 65 million years ago,
some 480,000 cubic miles (2 million cubic kilometers) of basaltic
lava poured across the land in less than a million years. Such
massive eruptions from Earth's interior are not taking place
anywhere on the planet today. If they were, it would be a sight
to behold.

According to vulcanologist Peter Hooper of Washington
State University, "The enormity of such an event is even hard to
imagine. Try to picture a lava front about as high as a five-story
building, more than 62 miles (100 kilometers) wide, and at a
temperature of 2,000°F (1,100°C) advancing down on you at 3
miles (5 kilometers) an hour."

When a plume rises into the upper mantle, it fans out as a
huge mushroom-shaped hot plate a few hundred miles across. It
then breeds volcanoes. Outside the region of a plume's hot, rising rock is a region of cooler mantle rock that is being sucked
down to replace the rising rock. In this way convection cells of
moving magma may be formed.

Geologists think that plate movement has been the rule for
at least the past billion years, possibly as long ago as four billion
years. But a lot more research is needed to gain a solid understanding of the forces that move the lithosphere's plates, and for
how long those forces have been at work.

Hot Spots and the Hawaiian Islands

An understanding of how a hot spot works may come by keeping a probing eye on the Hawaiian Islands. For it is there, in the
Pacific Ocean, where boiling magma has been burning its way
up through the Pacific Plate like periodic blasts from a welding
torch. The chain of some 125 volcanic islands and underwater

mountains that the Hawaiian group belongs to has been a relentless assembly line of hot-spot volcanoes for the past 65 million years.

The long chain of islands and flat-topped seamounts extends several thousand miles northwest from Hawaii, all the way to the Aleutian Trench, just south of Alaska. The chain of mountains that has been produced over millions of years as the Pacific Plate has been gliding over a hot spot now located under the volcanic mountain Mauna Loa. As the Pacific Plate has continued to glide across the hot spot fed by a gigantic plume, one volcanic island after another has been popped up out of the sea floor. Each volcanic island is a geologic clock that recorded the time when the volcano was perched atop the hot spot.

The oldest of these volcanic islands is a submerged group named the Emperor Seamounts chain. It extends southward in a straight line from the eastern tip of the Aleutian Islands. Their basaltic rock is about 65 million years old. Extending southeast from the islands is another straight line of mostly submerged volcanoes and seamounts called the Hawaiian Ridge. At the northwestern end of the ridge are the Midway Islands. Their volcanic rock is much younger, only about 27 million years old. The next visible island southwest of the Midway Islands is Kauai.

Kauai is the oldest of the large islands that make up the Hawaiian Islands. Its lavas are between four and six million years old. So six million years ago, when Kauai was over the hot spot, it was the only Hawaiian island. As the Pacific Plate kept gliding northwest, away from the hot spot, Kauai became inactive. A new volcanic island, Oahu, formed in its place over the hot spot and the plume that fed it. The lavas of Oahu are between two and three million years old. In its turn, Oahu, too, has glided off the hot spot toward the northwest, carried along by movement of the Pacific Plate. Next the island of Molokai formed. Its lavas are between one and two million years old. Then came Maui, whose lavas are even younger, less than a million years old. Today, the lava flows of Hawaii, the largest island of the chain, are fed by the ancient, but still very active, hot spot.

The old volcanic islands Kauai, Oahu, and Molokai have ancient lavas that have been deeply eroded. There are no new outpourings of lavas to feed those old volcanoes. They are dead. But the slopes of the island of Hawaii are paved with fresh lava flows. Two of its volcanoes, Mauna Loa and Kilauea, are very active and pour forth enormous amounts of new lava. They will probably remain active for another half million years or so. After that tick on the geologic clock, they too will slip off the hot spot that feeds them. We are made aware of the tremendous forces at work beneath the sea floor when we realize that the Hawaiian Islands, not Mount Everest, are our planet's mightiest mountains. They rise 33,465 feet (10,200 meters) above the ocean floor.

Geologists are keeping tuned to the Hawaiian hot spot in hopes of learning more about the forces that shape and reshape Earth's surface. But elsewhere there are about twenty others boiling away and reshaping the seafloor and the ground we walk on. They, too, will add to our knowledge about the marathon dance of the continents. As fascinating as that new knowledge is in explaining how Earth has come to be as it is today, an equally fascinating question is how those deep-Earth forces will shape the planet's future.

Eight

Earth's Future Face

As geologists probe planet Earth from the ground, astronomers study it from satellite observatories in space. Each year brings new information about the present state of the planet and its ancient past. The science of plate tectonics has clearly shown that the crystal ball holding secrets about Earth's future is the sea floor. For it is there where the planet is being torn apart along the mid-ocean ridges and pushed back together again along the deep-ocean trenches. As ocean floors widen, old lithospheric rock is forced down into the upper mantle, melted, and recycled again and again. And as hot spots boil up through the ocean floor, new land is formed.

Loihi: An Island in the Making

What is happening to the Hawaiian Islands today is giving us a small, local preview of an island in the making. About 20 miles (32 kilometers) southeast of the island of Hawaii, magma churning up out of the seafloor is building a future island. It is the Loihi

Seamount, now 8 miles (13 kilometers) wide by 16 miles (25 kilometers) long. Whether Loihi is being fed by the same hot spot that nourishes Mauna Loa and Kilauea is still a question. Regardless, another tropical volcanic island is being added to the Hawaiian Islands group. In the summer of 1996, more than three thousand earthquakes announced a new eruption of the seamount, which has built up to a height of 9,000 feet (2,749 meters) above the sea floor. That means it has another 3,000 feet (915 meters) to go before breaking out above the ocean surface.

Plate Movements and the Future

Plate movements have brought dramatic changes to the land and sea floor over the lifetime of our planet. Not only have they resculpted the face of the planet, but the shifting about has brought marked environmental changes and changes in climate. In turn, those changes have been an important driving force for the evolution of plants and animals ever since life began.

As the land and seas have changed in the past, they are changing now, and they will continue to change for billions of years into the future. Plant and animal fossils found in Antarctica show that that great land mass has not always been anchored at the South Pole, but at an earlier time was located closer to the Equator in a climate zone more favorable for palm trees than for penguins.

Based on what we now know about how the world's major plates are moving, geologists can make educated guesses about what Earth may look like some 50 million years from now. Antarctica may well stay pretty much where it is today, except it may twist a bit clockwise. Both the Atlantic and Indian oceans will broaden due to sea-floor spreading. At the same time the Pacific Ocean will be crowded smaller. Australia, which long ago

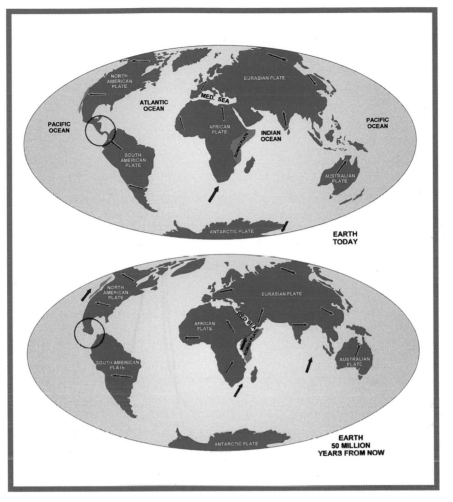

Geologists can make educated guesses about what Earth may look like some 50 million years from now on the basis of plate movement. While sea floor spreading broadens both the Atlantic and Indian oceans, the Pacific Ocean will shrink. Australia, which long ago broke away from Antarctica, will keep creeping northward and ram the Eurasian Plate. A large chunk of eastern Africa (orange) will split off along the great rift valley and slide northward. Meanwhile, Africa's general drift northward will widen the Red Sea and cram the Mediterranean Sea into a lake. In the United States, Baja California and part of California itself will break away along the San Andreas Fault and drift westward (orange). Although North and South America are linked today (orange circle), they may one day separate.

broke away from Antarctica, will continue its northward creep toward Southeast Asia until it rubs against the Eurasian Plate. A large chunk of eastern Africa will split off along the great rift valley and slide northeastward. Meanwhile, the general drift northward of the rest of Africa will block off the Bay of Biscay. It also will widen the Red Sea and cram the Mediterranean Sea into what will become the Mediterranean Lake.

In what today is the United States, Baja California and part of California itself will continue to break away along the San Andreas Fault and move westward into the Pacific Ocean. Some 10 million years from now Los Angeles will be up beside San Francisco, although it will still be attached to the mainland. Some 60 million years after that Los Angeles will begin to slide into the Aleutian Trench. Good-bye western California.

The overwhelming evidence for plate tectonics came in the late 1950s and early 1960s. It was the single most important idea in the history of geology. It explained so much that earlier had seemed obscure or uncertain—mountain building, the formation of ocean trenches, the cause of earthquakes and volcanoes, and certain climate changes. As the principle of evolution became the unifying concept for biology in the mid-1800s, plate tectonics has become the unifying concept for geology.

Glossary

Asthenosphere—that zone within the upper mantle where the rock is plastic and permits movement of the crust. It begins below a depth of about 60 miles (100 kilometers) and extends to a depth of about 450 miles (725 kilometers).

Basalt—a type of igneous rock. It is fine grained and dark gray to black.

Continental drift—the idea that the present continents once existed as a single supercontinent that broke into smaller continents, which then "drifted" to their present positions and continue to migrate.

Convection cell—within Earth's mantle, one of a pair of convection currents of molten rock that transports heat and magma up to the base of the lithosphere.

Core—the innermost region of Earth, surrounded by the mantle. There is a solid inner core and a liquid outer core of iron and nickel.

Crust—the outermost zone of Earth's surface. The continental crust is some 20 miles (35 kilometers) thick. The ocean floor crust is some 3 miles (5 kilometers) thick.

Fossils—ancient remains of plants and animals preserved in rock.

Geology—the science that examines Earth's composition and structure as revealed by its rocks. Geology also is concerned with Earth's origin and history.

Gondwanaland—an ancient continent formed during Earth's very early history, when a large supercontinent broke into a northern half and a southern half, the latter of which was Gondwanaland.

Hot spot—a region of melted rock that boils up to the base of the lithosphere and spills magma out onto the sea floor or onto the land.

Igneous rock—rock formed when molten material flows up from deeper parts of Earth's crust and solidifies either within the crust or at the surface.

Lava—volcanic outpourings of magma that cools and hardens to rock.

Lithosphere—Earth's rigid outer rock layer that extends to a depth of about 40 miles (65 kilometers).

Magma—fluid rock material originating in the deeper parts of Earth's crust. It is capable of forcing its way up through solid rock and, when flowing out over the surface, it is then called lava.

Mantle—that region of Earth's interior that lies between the outer boundary of the core and the lower boundary of the crust.

Plates—rock platforms that form Earth's crust. There are six major plates and about a dozen smaller ones. The continents, along with sections of the ocean floor, rest on plates that are pushed about like giant rafts of stone floating in a sea of puttylike rock. The edges of a plate are marked by intensive earthquake and volcanic activity.

Plate tectonics—the widely accepted notion that there are six major "plates" and about a dozen smaller ones that form Earth's crust. The continents, along with sections of the ocean floor, are pushed about like giant rafts of rock due to the movement of the puttylike rock of the asthenosphere below.

Plume—an upward flow of molten rock from the lower mantle to the crust, believed to form hot spots.

Rift valley—a fracture in Earth's crust along which molten rock from the mantle wells up and flows out onto the surrounding sea floor, or land.

Sea floor spreading—the widening of the ocean floor due to the upwelling of magma through ocean floor fracture lines that extend for hundreds of miles. The Mid-Atlantic Ridge is one such fracture line.

Sediments—loose bits and pieces of clay, mud, sand, gravel, lime, and other Earth materials that are washed into the oceans and lakes.

Smoker—mineral-rich water that boils up out of an opening in the sea floor and provides an environment rich in nutrients for a wide variety of ocean bottom organisms.

Strata—layers of sediments or rock.

Subduction zone—the region along which one plate collides with and descends beneath a neighboring plate. The depressed plate edge then melts.

Transformation fault—a fault, or crack in Earth's crust, along which two segments of lithosphere move against one another. The San Andreas Fault is an example.

Tsunami—a gigantic destructive wave triggered by an earthquake in the sea floor. Tsunamis may reach heights of 100 feet (30 meters) or more.

Uniformitarianism—the notion that past events in Earth's history can be explained by known laws and principles acting today.

Further Reading

Davidson, Keay, and A.R. Williams. "Under Our Skin." *National Geographic* (January 1996) pp. 100–111.

Deitz, Robert S., and John C. Holden. "The Breakup of Pangaea." *Scientific American* (October 1970) pp. 30–41.

Gallant, Roy A., and Christopher J. Schuberth. *Earth: The Making of a Planet*. Tarrytown, N.Y.: Marshall Cavendish, 1998.

Gallant, Roy A. *Restless Earth*. New York: Franklin Watts, 1986.

Gore, Rick. "Our Restless Planet Earth." *National Geographic* (August 1985) pp. 142–181.

Grove, Noel. "Volcanoes: Crucibles of Creation." *National Geographic* (December 1992) pp. 5–41.

Hallam, A. "Alfred Wegener and the Hypothesis of Continental Drift." *Scientific American* (February 1975) pp. 88–97.

Hecht, Jeff. "Tilt a World." *Earth* (June 1998) pp. 34–37.

Hurley, Patrick M. "The Confirmation of Continental Drift." *Scientific American* (May 1972) pp. 53–64.

Judge, Joseph. "A Buried Roman Town Gives Up Its Dead." *National Geographic* (December 1982) pp. 687–693.

Lindberg, David C. "Images of Earth in the Year 1000." *Earth* (December 1996) pp. 26–29.

Mestel, Rosie. "Mush in the Mantle." *Earth* (February 1997) pp. 22–25.

Moores, Eldridge. "The Story of Earth." *Earth* (December 1996) pp. 30–33.

National Geographic Society. Map: "Physical Earth" (May 1998).

Parks, Noreen. "Loihi Rumbles to Life." *Earth* (April 1997) pp. 42–49.

Pendick, Daniel. "Himalayan High Tension," *Earth* (October 1996) pp. 46–53.

Rona, Peter A. "Deep Sea Geysers of the Atlantic." *National Geographic* (October 1992) pp. 105–109.

Stager, Curt. "Africa's Great Rift." *National Geographic* (May 1990) pp. 2–41.

Weiss, Peter. "Land Before Time." *Earth* (February 1998) pp. 28–33.

Williams, Stanley N. "Double Trouble." *Earth* (August 1996) pp. 42–49.

Wysession, Michael E. "Journey to the Center of the Earth." *Earth* (December 1996) pp. 46–49.

Index

Page numbers for illustrations are in **boldface**.